THE MINERAL MANIACS
AND THE MAGIC HARDHAT

D1509101

written by JULES MILES
illustrated by MEG WHALEN

PAXTERRA
PRESS

Paxterra Press
Centennial, Colorado
www.PaxterraPress.com

Text Copyright © 2017 Julianne Miles
Illustrations Copyright © 2017 Meg Whalen
All rights reserved.

Cover designed by Meg Whalen
Graphics in the Digging Deeper chapter designed by Jonny Black
Paxterra Press logo designed by Gaby Seeley
Paxterra Press name and logo are trademarks of Paxterra Press
Photo on page 131 credit: NIOSH (public domain)
Photos of the gems on pages 134-136 are licensed through iStock.com
Photo on page 138 credit: Ron Landberg, BLM.gov (public domain)

ISBN-10: 0-9986985-1-2
ISBN-13: 978-0-9986985-1-9
Library of Congress Control Number: 2017906778

10 9 8 7 6 5 4 3 2 1
First Edition
Originally printed in the United States of America

For my mining engineer, who changed my life.

J.M.

For Danny, my rock.

M.W.

Chapters

The Hardhat

Everyone at Hillsdale Elementary knew Marabel, Victor, and Herbie. Best friends since their kindergarten days, the trio had become notorious for their silly and inventive pranks.

"Pull for Candy" signs on school fire alarms, Kool Aid packets taped to water fountains, and even drones dropping chocolate bars in the cafeteria. The trio's self-imposed name, "The Maniacs," had become synonymous for being goofy, clever, and almost always in detention.

On the afternoon of October 24, the maniacs were sitting quietly in afterschool detention in Room 13. Victor tap-tapped his pencil on the corner of his desk as he watched the hands of the clock move from one digit to the next. Herbie furiously scribbled in his notebook (which was always with him) as he let out a deep, bored sigh. Marabel, always the puzzle-solver, was busy with her newest crossword puzzle, which she finished only five minutes into the period.

And every so often, one of the maniacs would look up at their new detention teacher and smile. Ms. Pebbles, with her frizzy gray hair, large-rimmed glasses, and stained polo shirt, didn't seem to even notice the children were in the room at all.

"Umm, Ms. Pebbles?" Marabel spoke after ten minutes of silence, "shouldn't we, like, write 'I'm Sorry' on the chalkboard or something?"

Ms. Pebbles continued to look down at her desk, unaware of the girl's question or presence in her room. The teacher's face was inches from a large, purple textbook which laid flat on her desk. She swatted strands of her gray hair from her forehead as she sunk herself into her large book, every so often lowering her glasses and squinting her eyes.

Marabel looked at Herbie and Victor, both of whom seemed confused by the strange teacher's ambivalence. Marabel shrugged as she looked at her surroundings. Ms. Pebbles taught fifth grade science, but everyone at school simply knew her as "the Rock Lady," and her classroom demonstrated why.

Room 13 was covered floor to ceiling with posters of rocks, minerals, and mountains from around the world. A large white shelf circled the top of the entire room and contained samples of large gemstones. Three dark brown bookshelves sat in the back of the room and held books about every species of rock or mineral ever known. But catching the gaze of almost every student was the front corner of the room. There, in a dimly lit corner underneath a tall cabinet, stood a large wooden pedestal displaying a hardhat.

Marabel squinted her eyes as she stared at the large yellow hat. The hat was a plastic hardhat with a light on top, a lot like the hats worn by construction workers. She burrowed her brow as her curiosity peaked.

"Hey, Victor," Marabel whispered to her friend. Victor looked up from his desk as Marabel pointed towards the corner of the room. "I dare you to go steal that hat!"

Victor laughed under his breath. "No way! I don't wanna get another afternoon of detention! I'm missing my cartoon for this!"

Marabel rolled her eyes. "Whatever! Ms. Pebbles won't even notice; look at her!"

Victor looked over at their detention teacher. Ms. Pebbles was looking from one book to the next on her desk as she furiously copied notes into a small black notebook. Marabel smiled as she looked at Herbie, who seemed to be doing the same thing.

Victor bit his lip. He didn't want to miss another day of afterschool cartoons, but he really hated turning down a dare. Victor was always up for a challenge.

He looked towards his teacher than back at his friends. Herbie looked up as he realized his friends were plotting another scheme. He watched as Marabel egged Victor towards the hat in the corner of the room. Herbie typically liked a more behind-the-scenes approach to their pranks, but even he was getting bored by the day's detention.

"Go on, Victor! She's not watching!" Herbie whispered as he grabbed his friend's shirt and lifted him up.

Victor rubbed his hands through his short, brown hair as he slowly moved towards the hat. With every step he looked either towards his teacher or his friends, tip-toeing through the classroom as if on a mission. At one point, as a joke for his friends, he dropped to the ground in an army crawl. Marabel and Herbie giggled as Ms. Pebbles continued to seem unaware of her room's occupants.

Victor finally made his way to the front of the room and examined the yellow hardhat. He moved his fingers along the edges of the hat's brim as he looked back at his friends. Marabel nodded and smiled her approval.

Victor took one final glance at his teacher as he reached for the hat. "Hey Herbie, catch!"

Victor flung the hat towards his friend with all his might, sending the bright yellow helmet through the air. Marabel reached for it but missed, and instead the hat landed in the long arms of Herbie, who smiled and tossed it over to Marabel. Excitement filled the room as the hat flew from child to child.

"I guess you could do that if you'd like children," Ms. Pebbles spoke for the first time that detention, "but I'd be careful with that if I were you.

Marabel laughed as she snatched the hat from Herbie's hand. "Careful! With this old costume? What do you use this thing for anyway?"

Marabel smiled as she raised the hat to her head. In an instant, a bright light flashed through the room. A sweeping wind pushed its way through the chairs and desks. In the distance, the faint sound of a whistle echoed through the school's concrete walls, shaking the foundation and ripping paper from the walls. Victor and Herbie fell to the ground, covering their heads and shaking in fear.

As suddenly as it had begun, the chaos ended. Both boys jumped to their feet and looked around

the room. Ms. Pebbles shook her head as she carefully picked up posters from the floor.

"Wh—wh—where is Marabel?" Victor asked, shaking as he looked around the space.

Ms. Pebbles sighed as she laid down the broom. "She, I am afraid, has disappeared."

A Mysterious World

Marabel shivered and shuddered as she let out a soft sob. She was curled up in a ball on the floor of a dark, cold cave. She felt the wet soil under the tips of her fingers as she fearfully looked at her new surroundings.

The walls of the cave were made of rock, and the ground was moist soil and pebbles. The only thing that could be heard in the dark distance was the scurrying of feet and the loud echoes of a chorus of whistling.

Her heart raced as she tried to move her feet, but every part of her seemed frozen. "Victor! Herbie!" she whispered, inching in the direction of an open doorway.

"Ruby, is that you?"

Marabel fell back. Her voice quivered. "H—h—hello?"

"Ruby! Is that you? Thank goodness you're here!" The voice seemed to come from the hat itself, as if from a tiny speaker.

Just as Marabel reached for her hat to respond to the strange voice, a large stone object rolled past her, knocking the girl to the ground. Then there was another, and another. Marabel stared in disbelief as she realized they were rugged, rolling creatures made entirely of rock.

The rock formations were moving about as if they were alive, whistles coming from each creature as they scurried along the damp ground.

One after another, what appeared to be large rocks with arms and legs rushed past her, each whistling its own song as they sped by Marabel's frozen-in-place body.

Marabel's voice cracked, and tears streamed

down her face as she huddled in the corner. Where was she? Whose voice had Marabel just heard? And how could she get back to her friends? This predicament was much more puzzling than any of her brain teasers, logic puzzles, or word game books.

As she was about to call out again, one of the

larger rock people pushed past her with such force that she landed face-first on the ground, knocking the hardhat off her head and into her hands.

When she went to raise her head again, she was curled up in a ball, back on the tile floor of Room 13. Victor and Herbie wiped tears from their eyes as Ms. Pebbles patted Marabel's shoulder.

"Oh, good," Ms. Pebbles said calmly as she pried the hat from Marabel's grip and returned it to the pedestal. "I was worried that you might drop this. That would put us in quite the pickle."

Victor and Herbie shrieked with delight as they ran to welcome their returning friend, questions pouring out of their mouths at a mile a minute. Marabel remained frozen.

"Wh—wh—what happened?" Marabel asked, still trembling with confusion and fear.

"Wasn't the Paxterra world magical?" Ms. Pebbles asked with excitement.

Marabel barely blinked as she stared at her teacher, who seemed far too enthusiastic about the dangerous world Marabel just discovered.

"Paxterras?" Marabel repeated, as Victor and Herbie helped her to her feet.

"Yes, my dear, you visited the world of my treasured friends, the Paxterras. Quite a unique bunch, aren't they?"

Marabel started to get annoyed with her teacher's spunk. "Again I say . . . the Paxterras?"

Ms. Pebbles stared at the scared young girl and realized how bewildered Marabel was. She spoke calmly and clearly. "Yes, dear. Didn't you see the stone people?"

Marabel's eyes widened as she tried to comprehend what she thought she had just imagined. Everything she saw—the damp, muddy floor, coolness of the underground, and the people made of stone—everything was real. Her stomach turned a somersault as she gripped the corner of a table and quickly sat down.

"People made of *stone*?" Victor's voice broke Marabel's thoughts. The girl looked up at her friends, each with different but perplexed expressions. Herbie looked at Marabel in disbelief, as if he were waiting for his friend to correct the teacher's comments. Victor, on the other hand, had large, wide eyes that were filled with excitement and intrigue.

"Yes, people made of stone." Ms. Pebbles commented calmly. "They really are quite a fantastic bunch."

As Ms. Pebbles turned to her desk, a bright light steadily blinked from the corner of the room. One by one, as their heads turned toward the bewildering sight, the students slowly realized the light was coming from the hardhat itself.

Ms. Pebbles clapped her hands and dashed to her desk.

"Oh, my! They're summoning me again, and so quickly too! This can't be good!"

Ms. Pebbles rustled through papers on her desk. Aha! Here they are!" Ms. Pebbles exclaimed, as she held her large glasses in the air. She placed her spectacles on her face and headed toward the hat.

"You all will be okay by yourselves for a bit, will you not?" Ms. Pebbles asked as she grabbed the hat from its pedestal. "It appears I have places to be!"

"Wait! What?" Herbie's voice spoke out. "You're just going to leave us?"

"Are you going to the magical world place?" Victor asked with excitement.

"Yes, of course I am. They need me." Ms. Pebbles spoke in a hurried manner. She seemed annoyed by the children's continuous questions. "They are having quite a crisis, and if I don't go help them, soon *their crisis will also be our crisis!*"

Just as Ms. Pebbles was about to place the hardhat on her head, Marabel's voice pierced the air. "No! You can't go back! You can't just leave us here!"

Both boys looked at Marabel. They had never heard their friend so rattled.

The trio had known each other since the early days of Ms. Sue's kindergarten class. They bonded almost immediately over their mutual love of dirt, worms, and other ooey-gooey slimy things. And since that time, they remained fearless and inseparable, with Marabel almost always serving as the group's leader. Marabel thought of every prank, every class joke, and almost every risky scheme.

And yet here she stood, for what may have been the first time in her life, filled with fear and anxiety. She was used to being in charge and having control over the situations the Maniacs faced, but on the afternoon of October 24, Marabel had lost all control. She knew nothing of what lay ahead, and she didn't like it one bit.

Marabel looked at her friends then back at the teacher. She took a deep breath and tried to regain her composure.

"I mean, aren't we in detention?" Marabel asked reasonably. "Won't you get in trouble with Principal Hintz if you just leave us here?"

Ms. Pebbles lowered the hat to waist level and bit her lip. Her eyes moved back and forth as she tried to figure out the puzzle in her head. Finally, she sighed and again lifted the hat.

"I know, children, but believe me when I tell you that our world is dependent on me going. I trust that the principal would understand."

"The *world?*" Victor's curiosity had reached its peak.

"Yes, dear children, the *world*. Now if you don't mind, I'll be off."

Marabel, Herbie, and Victor let out shrieks as they collapsed to the floor and braced themselves for whatever would happen next. Ms. Pebbles looked around the room one last time, took a deep breath, and gently placed the hat on her head.

The Mission

Victor slowly moved his hands from his face. His eyes gazed at the corner of the room.

"Whoa! You're still here!"

Marabel and Herbie quickly lifted their heads and stood up. They looked in the corner of the room where a frazzled Ms. Pebbles stood.

"Well, that's awfully strange," Ms. Pebbles said, as she grabbed the hat from her head and examined it. "It still appears to be blinking, so I know it's working. Let me try it again."

The kids gasped as Ms. Pebbles lifted the hat and placed it on her head. The room stood silent for a minute while each person stared at one another with confused expressions. Ms. Pebbles seemed the most astonished of all.

"Well, this just isn't right," Ms. Pebbles said, shaking her head. Her fingers fidgeted as she quickly paced from one corner of the room to another and muttered under her breath. Victor and Herbie giggled as they watched their detention teacher wave the hat in the air as if she were looking for cell phone reception. Finally, Ms. Pebbles had an idea.

"Yes, the phone! I shall use the phone!"

Ms. Pebbles rushed to a cabinet directly above the hardhat pedestal. She rose to her tiptoes and pulled down an old-fashioned telephone receiver and pressed a switch located among the cables and wires of the phone box hidden within the cabinet.

"Whoa! That thing is huge!" Victor shouted as he got a closer look at the unknown object. Ms. Pebbles placed her fingers in different numbered holes and turned the dial. The children gasped as they heard the click-click sound of the phone as the dial slowly rotated back to the center. Ms. Pebbles lifted the phone to her ear and waited.

"Yes, hello, King. The hat does not seem to be working," she spoke clearly into the phone's receiver.

Marabel and Victor's eyes widened in awe. King? Ms. Pebbles was talking to a *king*?

"Yes, I tried it a couple of times, and it is still blinking." Ms. Pebbles coiled the wire of the phone around her long, bony fingers.

"No, that wasn't me," she chuckled. A little person I am with took the hat by mistake."

Little person? Marabel wondered. She gulped as she realized the teacher was talking about her. Normally, Marabel hated being called little, especially by grown-ups. She hated having to listen to every rule and every order from grown-ups, simply because she was smaller.

But in that moment, Marabel did feel little. She wanted to shrink and shrink and keep shrinking until her world made sense again.

Herbie squeezed Marabel's hand. He smiled at

his friend, trying to reassure her that everything would be okay, but Herbie wasn't sure he even believed that anymore. The threesome had been in trouble many times in their young lives. But trouble with a king from a magical world of stone people? That was not the kind of trouble Herbie was prepared to deal with.

Victor strode purposefully to his teacher. He leaned his body close to Ms. Pebbles' shoulder as he quietly tried to listen to the conversation.

Ms. Pebbles noticed the meddlesome boy and swatted Victor away. Her body tensed and her voice raised as she spoke into the receiver.

"No, King, you are mistaken." Ms. Pebbles paused as she listened to further instructions. She nervously shook her head. "No, King, that is simply not possible. No. No! They are just *children*, King. It simply cannot be done."

Victor quickly looked toward his friends then back at the teacher. Was Ms. Pebbles talking about them?

"Yes, of course I know what is at stake, but I cannot allow that!" Ms. Pebbles wailed into the phone.

Herbie looked up from his desk startled. He had never seen the Rock Lady so flustered.

Ms. Pebble's face turned pale and resigned as she looked at the blinking hat.

"OK, I understand," Ms. Pebbles quietly agreed into the phone's receiver. "I will see what I can do."

Ms. Pebbles gently placed the phone back in the cabinet. She swatted strands of her hair from her forehead and looked at Marabel.

"Well, dear girl, it appears you have meddled with the magic to a point that now you must fix it."

Ms. Pebbles grabbed the blinking hat and moved toward the students' desks.

Marabel fought back tears and frantically shook her head. "Please, please don't make me go back!"

"I'm sorry, Marabel, but this is your mistake to fix." Ms. Pebbles placed the hat on Marabel's desk.

Marabel wiped the tears falling down her cheeks and continued shaking her head. Herbie grabbed Marabel's hand and prepared to defend his friend.

"Wait, Ms. Pebbles!" Herbie protested. "What mistake exactly does she need to fix? You haven't even told us what's going on!"

Agreeing with Herbie, Marabel wiped her cheeks and stared at the teacher. Ms. Pebbles nodded.

"Children, there is much to explain," Ms. Pebbles said gently, "but we simply do not have the time right now. This is an urgent matter which must be handled immediately, and for some reason, I can't be the one to complete it."

Herbie and Marabel looked at one another. Herbie shook his head.

"Ms. Pebbles, you're asking us to do something we know nothing about!" Herbie insisted. "That just seems crazy to me!"

Ms. Pebbles chuckled. "You're right, Herbie. I am asking you to trust a teacher you barely know, to help a magical king you've never met, to go on an adventure which could put you in grave danger. What I am asking sounds silly and nonsensical. But I am asking you to have a bit of faith and perhaps a spirit of adventure too."

"I'll do it!" Victor yelled from the front of the room Marabel, Herbie, and Ms. Pebbles looked at Victor as he grinned and ran eagerly toward the hat.

"I wanna go! It's not every day you get to visit a

magical world!"

Victor reached for the hat, but Ms. Pebbles quickly swatted his hand away.

"You won't be going to the magical world, Victor."

"Please! Oh, please!" Victor pleaded.

Ms. Pebbles shook her head. "No, you don't understand. The hat won't be taking you to the magical world of the Paxterras. It has other plans in mind."

The children looked at each other then back at their teacher. "What do you mean?" Herbie asked.

Ms. Pebbles gently stroked the edges of the hat with her finger. Her face became serious and stern. "Trust me; I simply don't have enough time to explain. But what I can tell you briefly is this: If we don't act soon, the entire supply of a certain rock or mineral will be gone from this world . . . forever!"

Blink-blink. Victor, Herbie, and Marabel stared blankly at their teacher. A rock? This whole far-fetched plan is to save some silly rock?

"So what?" Victor asked, confused. "Who needs a bunch of old, dirty rocks?"

Ms. Pebbles' eyes widened. "Oh, children! You don't seem to understand! Our world is made of rocks and minerals!"

Marabel squinted her eyes as she tried to understand her teacher's words. She had been on the earth for a whole ten years, and if there's one thing she knew, her world was definitely not made of a bunch of rocks. Or was it?

"What do you mean, Ms. Pebbles?" Herbie's voice pierced through Marabel's confusion.

Ms. Pebbles sighed. "Children, everything you have come to rely on in your world—from the cars you ride in, to the phones you use, to the buildings where you live—is dependent on the resources of our earth. Believe it or not, almost everything we own and use is made of rocks and minerals."

"No way!" Victor shouted. "If that is true, I have to go! I need to see some proof!"

The teacher glanced at Marabel as the young girl

rested her head in her hands.

"OK, Victor," Ms. Pebbles said. "If Marabel can't fix this blunder, then I suppose you must."

Herbie and Marabel jumped to their feet. Marabel grabbed Victor's hand and yanked at his arm. "Please, Victor, don't do it. You know it's dangerous!"

Victor smiled as he squeezed his friend's hand. "That's what makes it so exciting, Marabel! A real-life adventure!"

Ms. Pebbles grabbed Victor's hand and led him to the front of the room.

"All right, Victor. I do not know where the hat will send you, but it is somewhere in our earthly world. While you are there, you will be completely invisible to anyone you may encounter, but please be careful! They won't be able to see or hear you, but if you rustle the ground too much you could draw attention to yourself!"

Victor nodded his head.

Ms. Pebbles continued. "I will not be able to

have contact with you during this time. The king will communicate to you through the hardhat and will give you instructions on what to look for. You must be very observant, dear boy, and try your best to recall everything you see. You may go to only one place, or you may go to several. I simply do not know."

"I'll do my best, I promise!" Victor said with confidence.

Ms. Pebbles smiled as she rested her hands on Victor's shoulders. "Yes, I trust that you will, dear boy. But one final thing: Never, under any circumstances, let the hardhat touch the ground without you holding it. Doing so will trap you at your location."

Marabel let out a soft whimper as she choked back tears.

Victor nodded again and took the hat from Ms. Pebbles' hands. He looked at his friends with a forced smile. He gave a thumbs-up as he lifted the hat to his head and, in an instant, disappeared.

The Heat

Victor landed with both feet on a soft, plush landscape. He barely had time to catch his breath when he noticed he was in a forest, surrounded on all sides by huge trees which seemed to reach the sky. The ground was moist and lush, and sounds of birds encircled him.

"Are you there?"

Victor jumped backward, tripping over his own feet and falling down with a loud THUMP! He looked around as he tried to find who belonged to the voice.

"Hello? Hello? Who am I speaking with?" The voice emerged in a soft, low tone. Amazed, Victor realized it was coming from the hat itself.

"I'm Victor," the boy whispered as he stared at the tip-tops of the trees surrounding him. Are you the mysterious, mighty, hardhat stone king . . . ?"

The voice from the hat laughed. "I am King Sapphire, King of the Paxterras, and the only one who can guide you through this risky mission. Thank you for taking on such an important task."

Victor stood tall and smiled. He was proud of his bravery. "No sweat, King Man!"

"OK, little one, we must begin," the voice echoed through the hardhat. "Could you please tell me . . . .?"

"Hey!" Victor interrupted. "Who you calling little?"

There was a pause from the hat. "Yes, sorry, Victor. I suppose human beings do not like to be referred to as children, even though . . . you are one. No matter, we Paxterras don't care about such things."

"Hey, do you want my help or not?" Victor

realized the king was making fun of him.

"Yes, of course. Sorry, little . . . I mean brave warrior, Victor."

"That's better!" Victor looked around the forest. "So, what am I here for anyway?"

"You will learn all in good time, Victor," declared the king, "but right now you must tell me what you see."

"But are you all in some sort of trouble? What's your world like? How'd you get to be king? What do you guys eat?"

"Enough!" boomed the king's voice in the hat. "Are all human children so nosy and impatient? We haven't the time for your questions right now! Please, begin describing anything you see!"

Victor sighed. "OK, OK."

Victor took a deep breath as he slowly tried to move. Victor oomped and whooshed as he pumped his arms, but his legs seemed glued to the ground beneath. It didn't take long for him to discover why.

"Hey! There's huge boots on my feet!" Victor

exclaimed. "I can barely walk in these things!" Victor grabbed his thigh as he tried to lift his legs through the muddy soil.

"Ah, how nice to hear!" the king's voice echoed. "The hat must want you to stay as safe as possible."

"What do you mean?" Victor whispered.

"Well, if you are going to a place with construction equipment and other important machinery, you need steel-toed boots and a hardhat! It is the very least you can do to stay safe. I am surprised, however, that the hat didn't also give you a safety vest and glasses. Those are pretty standard in mines."

Victor gasped as he grabbed his shirt and reached for his eyes. "Aw, man! I have those things, too!"

"Stop your complaining, Victor!" the king said angrily. "Safety must always come first! The hat knows this, and now you do, too. But no more waiting. You must figure out where you are!"

"Fine," Victor mumbled as he grumpily looked

to his feet.

Victor inched step by step through the soft, muddy earth beneath his feet. The chirp-chirp of the birds filled the sky as the young boy looked left, right, up, down.

"I'm in some kind of forest," Victor reported softly.

"No need to whisper, Victor." The king's voice echoed through the hat. "Remember, no one can see or hear you. Just don't make a lot of racket, and you should be just fine."

"So, what am I supposed to do?" Victor asked as he slowly stepped through coarse, woody debris.

"We must figure out where you are," the hat spoke into Victor's ear. "Describe your surroundings a bit."

"Well, it's a forest," Victor spoke in a low voice, still nervous about the possibility of being heard. "It's really wet. And it's kind of noisy. There are birds everywhere, and they keep squawking."

Victor looked at some of the birds circling over

him and for a moment was blinded by the sun's rays through the tall trees.

"Interesting," the king commented, softening his tone. "You must be in a rainforest. Tell me, is the climate rather hot?"

"Yeah! Now that you mention it!" Victor wiped away sweat that was dripping down his forehead.

"What else do you see?" the king asked. "Use all of your senses, Victor, but don't dillydally! You only have a few minutes!"

Victor quickened his pace through the dense forest. Suddenly, a loud noise stopped him in his tracks. He looked through the trees and saw large equipment, much like the kind found at construction areas. Victor knew immediately what they were.

"Oh, cool! I see an excavator, a bulldozer, and several huge dump trucks!"

"What's that, boy?"

"It looks like a construction site of some kind," Victor repeated. "Just like the boots predicted! I see a lot of cool machines and a *really* deep hole in the ground. There are ridges on the side of the mountain where it looks like a road for heavy-duty haul trucks. This is awesome!"

Victor was so excited to be the first of his friends to experience this thrill!

"Oh, wonderful!" the king proclaimed. "You must be at a mine. That's splendid! This might be easier than we thought."

"Easier? Easier than what?" Victor asked as his feet inched along the sandy, brown soil near the top of the mine.

"Never mind. Just keep telling me what you see."

Victor looked toward the mine and scratched his forehead. "Wait, I always thought a mine was below the ground."

"Yes, Victor, some mines are in caves or caverns, while others are built into ridges of mountains and dug out from the surface of the earth. There are many varieties, you see! This bit of information will help us immensely!"

Victor smiled. He was happy to please the king, even if he had absolutely no idea what was going on or why he was on the expedition in the first place.

Just as he was about to step foot onto the sandy, brown surface at the top of the mine, Victor's head

started to heat up. For a moment, he thought that the climate was a lot hotter than he had originally noticed. But this was different. The heat seemed to be coming from . . . the hardhat.

"Umm, Mr. King?" Victor called nervously into

the air. "I think the hat is on fire or something. It's burning my head like crazy!"

"Oh, no!" the king screamed frantically. "Victor, brace yourself!"

The Heights

Victor closed his eyes and firmly held the hardhat as light flashed all around him. When he opened his eyes, he was no longer in the bright warmth of the rainforest.

The ground beneath his feet was hard and dry. A cool breeze and the blue sky's brightness made Victor gasp. He realized that beautiful peaks and rough mountain ridges were the only sights for miles.

"I'm on a mountain now!" Victor exclaimed. "It's nothing like where I was before."

There was silence on the other end. Victor was worried that he lost contact with the king when the hardhat transported him to the new location.

"Hello? Hello? Are you there?" Victor nervously screamed into the mountain air.

"Yes, yes, Victor. I'm here." The king uttered in confusion. "I'm just trying to make sense of this new information. The two climates you described are very different from one another. It seems a bit strange to me, that's all."

"Why? Ms. Pebbles said I might go anywhere in the world!" Victor recounted.

"Yes, that's true," sighed the king, "but typically what I am looking for on these journeys occurs in climates similar to one another. But never mind that! Just be observant and tell me what you see!"

Victor started to hike up the ridge to get a better view. Suddenly, another startling sound made him fall face-first onto the rocky ground.

"There's someone up here," Victor whispered, now slithering silently across the rough terrain.

"Oh, how interesting!" The king said with a bit more excitement than Victor wanted to hear.

"Could you get a closer look at your company? Don't disturb your surroundings too much. They might think you're a wild animal. And that would be a bit of an unpleasant situation for everyone."

"Wait, why?" Victor asked, his nose pressed to the ground.

"Well, sometimes prospectors and highly skilled miners like to dig for rocks and find dinner all in one trip. It really does save time, you see."

Victor gulped as he realized the king's point. "Meaning . . . I would be dinner?"

"Not a very tasty one!" The king laughed through the hat.

Victor frowned. He did not appreciate the stone king's sense of humor.

"Quickly, now! Forget my jokes! What do you see?"

Victor stood up and climbed the ridge. He inched his nose over the top of the rock formation to get a closer look at the people who were working about twenty feet away.

Three men sat next to a large wall of the mountain. They were probably around Ms. Pebbles' age. Each of the men held small metal tools that they used to hammer and dig into the side of the mountain.

Victor was about to relay his findings to the king when he noticed a large piece of wood sticking out of the ground, about ten feet from the prospectors. The wood had a piece of cloth at the top, like a flag, and some writing on the side. Victor knew he had to get closer to read it.

"There are three men digging for rocks," Victor whispered. "It doesn't really seem like anything special, but there is a wooden stake sticking out of the ground right by the men. It has something written on it that I can't quite make out."

"Interesting," the king's voice vibrated through the hat. "Make sure you tell Ms. Pebbles about that when you return. There is something strangely human about that bit of information!"

As Victor looked over the ridge toward the men,

he heard them shout. The sound made Victor hit the dirt again. He was afraid that he would soon be a mountain man's dinner.

Victor peeked over the ridge and was surprised to see the men high-fiving and congratulating one another. One of the men held up a large gemstone in his fingers. It was difficult to see as the sun's rays glared down, but Victor noticed blue tints underneath the specks of dirt.

"I think the guys just found a really cool rock," Victor informed the king. "It looks pretty dirty, but I can see some blue parts to it."

"Yes! Very good, Victor!" the king exclaimed. "This is just the sort of information we need! Keep looking around; perhaps you could get a closer look at that blue stone."

Victor looked over the gray peaks as the sun streamed on his face. The last time Victor had seen such a beautiful mountain pass was in a picture on his dad's desk.

Victor sighed. Until the end of last year, Victor

had spent almost every weekend going sightseeing with his dad. They visited museums, went to the tallest peaks around their town, and even trod into rivers looking for cool rocks and fish.

But it had been almost six months since their last outing. His dad had started working more and just didn't seem to have the time for great adventures. Victor wished his dad was with him to experience the beauty of the majestic peak.

Victor started to head back to his original spot. He walked slowly and carefully, examining everything beneath his feet in the hope of finding another sample of the mysterious blue gemstone. Just as he was about to give up, a strange sight caught his eye.

A dark hole formed shadows on the side of a large, rocky cliff. Victor's eyes widened as he realized what he had just stumbled upon: a cave!

At the entrance of the cave were two old, wooden signs which read "DO NOT ENTER" and "ABANDONED MINE." Victor smiled from ear to ear as he realized his journey was about to get even more exciting.

"It's an old mine!" Victor said excitedly to the king.

"What? What are you talking about?" the king asked.

"I see a cave down at the bottom of this ridge. There are signs that say it's an abandoned mine. I'll check it out. How exciting!" Victor ran toward the cave.

"Do not go into that cave, Victor!" the king's voice said sternly inside the hardhat. "You must never, ever, under any circumstance, go into an abandoned mine. It is very dangerous!"

"Yeah, yeah," Victor shrugged as he ran to the entrance of the cave. Clearly the king didn't understand Victor's expertise in matters of adventure. Plus, Victor thought, what was the harm in going in just a couple of feet?

"Victor! Victor, you come back, right this instant!" Victor continued to ignore the king's command. Even the fate of the world had to take a backseat to exploring a cave!

Victor stepped into the cavern and looked around. He rested his hand on the side of the cave's entrance. The cave was dark and damp, and Victor began to worry that a swarm of bats would swoop into his path at any minute. It started gradually, but just as Victor turned to exit the cave, patches of dirt began to fall from the ceiling. The sides of the cavern began to shake, and larger rocks were soon falling in rapid speed.

Victor looked at the entrance as he realized what was happening: The old mine was caving in.

A Mine Claim

In a moment of quick thinking, Victor did the single thing that one should never do when heavy objects are falling: He removed his hardhat. In an instant, he was transported back to Room 13, surrounded by the quiet mumblings of his friends and teacher.

Marabel, Herbie, and Ms. Pebbles gasped as Victor emerged from nothingness into the corner of the classroom. During his absence, Ms. Pebbles had spent her time pouring through the large textbooks on her desk. Whenever Herbie or Marabel had tried asking about the Paxterras or their missions, Ms. Pebbles would simply reply, "All in good time, children."

The three ran to Victor, who was also breathing a sigh of relief at his return to familiar surroundings. Ms. Pebbles grabbed the hardhat from Victor's hands. She gently dusted the surface of the hat as she rubbed her sleeve around the hat's brim.

Herbie clenched his pencil and notebook in his hands as questions poured out of his mouth. "What did you see? Who did you meet? What was your mission? Did you find a stone person?"

Marabel grasped Victor's hand and squeezed tightly. She felt conflicted and scared: On the one hand, she was a puzzle solver. She wanted to discover what was going on and how to fix it. But on the other hand, she had no desire to ever put the hardhat on again, and she certainly didn't want to end up in the other world.

Victor smiled and raised his hands to the air, shushing Herbie and removing Marabel's hand from his. He smiled confidently as he walked to the chalkboard and handed Ms. Pebbles a piece of chalk from the holder.

"OK, Ms. Pebbles," Victor said calmly. "I have no idea what's going on, but here's everything that happened."

Ms. Pebbles began writing everything on the board, word for word, that Victor said. Victor's excitement and volume rose with each sentence.

"Oh, yeah!" Victor spoke confidently. "There was this stake or flag thing on top of the mountain. The king said you might know what that was all about."

Ms. Pebbles nodded. "Oh, yes, if there were prospectors, then you were probably on a mine claim."

"A mine claim?" the children asked simultaneously.

Ms. Pebbles turned to them. "Why, yes! Don't you know what a mine claim is?" she asked with surprise.

The children shook their heads.

Ms. Pebbles rolled her eyes. "Goodness me!" the

teacher's voice shrieked. What are they teaching you in school nowadays?"

Marabel smiled. "You mean, what are *you* teaching us in school nowadays?"

Ms. Pebbles paused as she rested the tip of the chalk on the board. "Well, yes, I suppose that is true. No matter, no matter. A mine claim is a piece of land owned by individuals for the sake of mining. They can't build on the land, but they can dig its minerals with the proper permits."

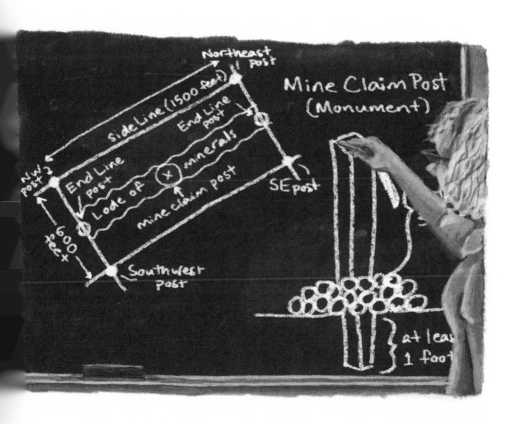

Ms. Pebbles began drawing a small mine stake on the board. "Any rock they find on the land," she continued, "or under the land, is basically theirs. There are about a dozen rules about claims which we just don't have time to cover today, though it is one of my favorite discussion topics."

Victor's eyes widened. He never knew that people could own whole pieces of mountains! As he opened his mouth to ask more questions, Ms. Pebbles abruptly turned and faced him.

"Oh, Victor!" she questioned fearfully. "You didn't happen to touch anything or take anything from the mountain, did you?"

Victor quickly shook his head.

Ms. Pebbles relaxed and breathed a sigh of relief. "Oh, thank goodness!" she said as she continued writing. "It is very disrespectful and *very illegal* to dig or mine on another person's mine claim. OK, Victor, keep telling me what you saw."

Victor recounted to the group some of the details he noticed on his journey. In the first location, the area was a tropical climate, but the mine looked like a giant dirt pit. The soil appeared to be reddish and clay-like. It was difficult to tell what they were pulling out of the red earth, but it was a very extensive operation, with dozens of machines, trucks, and large cylinder-shaped buildings in the background.

The second location was not nearly as busy, but it was impossible for Victor to figure out exactly what mineral could be mined on the mountain. The only detail Victor remembered about the stone discovered by the prospectors was that it had some green or blue tints.

As Victor finished his story, Ms. Pebbles sighed. The children looked at each other, then back at their teacher. What was going on? Why did Ms. Pebbles seem so frustrated?

"Ms. Pebbles," Marabel began. "Is there any chance you could tell us what's going on?"

Ms. Pebbles looked at the young girl perplexed. The students were only children, after all. What could they do to help the situation? Then again, the magic of the hardhat seemed to think they could help, so who was Ms. Pebbles to challenge the magic?

Ms. Pebbles sat down at her desk. She decided that it was only right to explain to the children what was happening. But just as she began the very long and complicated story, a strange sight caught her attention.

Continuous Blinks

Ms. Pebbles and the children jumped as they all stared at the same view: The hardhat was blinking again. Ms. Pebbles ran to the hat and quickly removed it from its pedestal.

"How bizarre!" she mumbled as if she were alone in the room. "Why would the hat need me now? I haven't figured out the puzzle!"

Puzzle? Marabel's eyes widened. Ever since she was a little girl, Marabel loved puzzles. Jigsaw puzzles, crossword puzzles, she loved them all! Marabel smiled as she remembered how much her mom loved telling people that she completed her first 100-piece jigsaw puzzle at just two years old.

But her smiled quickly disappeared as she realized she faced a puzzle she could not solve. More importantly, she didn't think she wanted to solve it, especially if it meant going back to the Paxterra world.

"Do you need to go, Ms. Pebbles?" Marabel asked her teacher.

Ms. Pebbles nodded. "Victor, thank you for your information. I will do the best I can with it, but I think my friends need me!"

Victor and Herbie protested. "But we wanna know what's going on!" they both echoed.

Ms. Pebbles shook her head. "No, not yet boys. I must go!"

Everyone braced for more chaos as Ms. Pebbles lifted the hat to her head. Marabel closed her eyes, not willing to watch yet another person disappear.

Several seconds went by before Marabel opened her eyes again. She was surprised to find Ms. Pebbles standing before them with an annoyed expression.

"Oh, not this again!" Ms. Pebbles was flustered. "Why is this hardhat not working?" She turned to face Victor. "Young man, are you sure you did everything you were supposed to do? Maybe you missed a location?"

Victor scratched his head. "I don't know," Victor said with confusion. "I just did what that king in the hat told me to do. But I'll gladly go again if you want me to!"

Before Ms. Pebbles could answer, Victor snatched the hat from her hands and quickly placed it on his head. Ms. Pebbles protested as she tried to grab the hat back, but everyone stood still when Victor, too, remained in the classroom.

Victor tap-tapped the side of the hardhat as it rested on his head. "Hello?" Victor shouted. "Hello! What's going on, King Sapphire?"

Marabel's heart raced. Did the hat want her?

"Do you think it's for Marabel?" Herbie asked as he looked at his friend. Marabel's face went white.

Ms. Pebbles shook her head. "No, no. This just

can't be!" She began quickly pacing back and forth across the room, flailing her arms and stomping her feet. "Just one mission is normally enough! What could that silly king be thinking?"

Herbie knew his question had worried Marabel, who hadn't spoken in several minutes and looked as if she were about to pass out. He watched as his other best friend tap-tapped on the side of the hat, begging to be sent on another mission. And in an instant, Herbie knew what must happen.

Herbie clutched the corners of his notebook as he breathed deeply and slowly. He patted the pencil tucked in his back pocket and walked up to Victor.

"Victor, I think it's my turn," Herbie announced as he held out an empty hand.

"No way!" Victor protested. "I'm the one who's supposed to go on adventures!"

Herbie nodded. "I know, Victor," he reassured his friend, "but for some reason, the hat doesn't want you to go now. It must want me."

Victor nodded his head and made a pouty face toward his friend. "Yeah, whatever. You could try, I guess," he said bitterly, as he handed the hat to Herbie.

Ms. Pebbles and Marabel glanced at the boys. Marabel stood to protest but quickly closed her mouth. Her fear was overwhelming, so she sat back down to let the others decide who would go.

Ms. Pebbles approached Herbie and placed her hands on the young boy's shoulders.

"Herbie, my dear," Ms. Pebbles spoke calmly, "I must be completely honest with you. I have absolutely no idea what is going on. But I suppose life can be exciting in that way!"

Herbie squinted his eyes and gulped. He was not amused by his teacher's optimism.

"OK, remember what I told Victor. Record everything you see, Herbie. Your notebook will come in handy!"

Herbie looked down at the content of his hands. His most trusted notebook companion in one hand;

the hardhat, the root of all his anxiety and confusion, in the other. He grasped both objects close to his chest as he stepped away from his friends. With a final wink toward Marabel and Victor, Herbie lifted the magical hardhat to his head and, in an instant, disappeared.

Manufactured Success

Herbie clutched his notebook and opened his eyes. He looked up and down as he squinted to become acquainted with his surroundings.

The fluorescent lights of the room gleamed off metal machinery and walls. The floor was a cold tile surface, much like the classroom, but the room was much louder. There were large machines around him, as well as men and women wearing hardhats and goggles.

Herbie prepared to speak to the mysterious hardhat when he noticed something strange. His hands quickly grabbed at his face as he realized that there were protective shields on the sides of his glasses.

He quickly remembered Victor's story about the boots appearing on his feet at the mine site. However, unlike Victor, Herbie welcome the small inconvenience of the shields. He smiled at the discovery that the magic of the hardhat seemed to care so much about his safety.

"I'm in a factory or something, said Herbie quietly into the air. He waited, anticipating the hat to speak, just as it had for Marabel and Victor.

"Ah! Another human person!" the king transmitted from the hat's brim. "How interesting!"

Herbie jumped at the deep sound of the king's voice. He cleared his throat as he stood tall and brave.

"What do you mean?" Herbie asked the hat.

"Well," the king began, "the magic of the hardhat is usually reserved for one human person: your teacher, Ms. Pebbles. But it is acting rather peculiarly today! Apparently, you must each have some kind of quality which our world desperately needs."

Herbie was astonished. "A special quality? Me?" He had never really thought about having any special qualities before.

"I'm sure you will be very helpful," the king said kindly. "To the task at hand! What do you see? Describe your surroundings!"

Herbie rushed to one side of the factory and pulled the pencil from his pocket. He feverishly drew pictures in his book and made notes of everything he saw.

There were large tanks and pipes spread throughout the space, possibly like the ones Victor had seen at the mine. But upon closer observation, Herbie noticed these machines had a temperature dial at the bottom.

"I think I'm in a room of furnaces or something," Herbie said as he wiped his forehead. "It's crazy hot in here!"

"Yes, that's interesting," the king commented. "Sometimes in the manufacturing process, minerals are heated to extreme temperatures in order to be purified, and from there they can be cut and shaped for various purposes. Try to move to another location to see if you can get a better look at the material."

Herbie looked around and noticed a door to another room. A couple of the workers were going in and out of the area, which meant the other room might have more clues. Herbie waited for a worker to open the door and then quickly scurried past him, almost brushing the side of the worker's coat.

"Phew, that was close!" Herbie said to himself. He looked at the new space, which was as well-lit but not nearly as noisy or hot as the furnace room. A group of workers in the corner hovered around a table. Herbie walked near the work space and realized the workers were about to cut into a large, clear material.

The words "CLEAR." "ROUND." "THICK." "LOOKS HARD" were all scribbled onto his pad just as the men lowered the large object onto a board. Herbie desperately wanted to see how the material was cut, but he knew he probably should find out whether there were any other clues about the kind of factory he occupied.

Herbie tip-toed his way through the narrow, cold room. Just as he reached for the handle to make his escape, a small hole in his shirt got caught on the edge of a table. He grabbed at the corner of the table, catching himself but knocking over a small, transparent material in the process. The clear, small cylinder, no wider than a watch, fell on the tile floor, making a faint "tink."

Herbie held his breath as he looked at the men.

"What was that?" one of the workers asked as he slowly walked in Herbie's direction.

Herbie didn't move a muscle. He heard the frantic questions of the king echoing in his head, but he didn't pay attention. All he knew in that moment was that he had to stay still—perfectly still.

"Ah, you know how a breeze from that door is always knocking stuff off tables," muttered another

worker. "Just go pick it up and make sure it didn't get scratched."

The first worker walked toward Herbie and gently picked the small disc off the floor. He brushed it with his sleeve and blew the dust off its surface. "No scratches! It's fine!" the worker said as he joined the rest of his company.

Herbie breathed a sigh of relief. He looked at the material and wrote "VERY HARD TO SCRATCH" in his journal.

"Young boy? Are you alright?" Herbie was finally able to focus on the king's questions.

"Yes, I'm fine," Herbie answered, "and my name is Herbie, by the way."

"Yes, yes, I'm so terrible at introductions," the king muttered quickly. "I appreciate that you aren't nearly as talkative as the last young person I spoke with."

Herbie smiled. Victor really was quite a talker.

Herbie looked at the factory and listened to the pitter-pat, screech-screech of the machinery. To

Herbie, it sounded beautiful. Ever since he was little, Herbie was fascinated by machinery and factories. Perhaps the magic of the hardhat knew this about him?

Before he could investigate the factory more thoroughly, a piercing heat filled his scalp.

"Ouch! This magic hurts!" Herbie hissed.

"Oh, dear! That wasn't much time!" the king said frantically. "Brace yourself, Herbie!"

A Gem of a Story

Herbie held his journal tightly as a flash of light filled the factory. He closed his eyes and held his breath. He knew the hardhat wasn't quite done with him yet.

When the heat in the hat cooled and the wind simmered, Herbie opened his eyes. The room was small, quiet, and still.

An older gentleman was sitting at a desk which held dozens of large stones, metal tools, and utensils.

"I'm in a small room with one other person," Herbie whispered, afraid to move. "I think he's cleaning rocks or something at his desk."

"Oh, that's wonderful!" the king echoed happily into the air. "I just love hearing of humans taking such care of our precious stones. Not many humans live this way, you know."

Herbie rubbed his forehead. He had only been in the room about two minutes, but for some reason, heat started to radiate from the hat again.

"Umm, Mr. King?" Herbie asked as he tried to fan away the heat. "My head is, like, on fire."

"Oh, no!" the king shouted. "So soon? Quick, Herbie, you must look for any clues you can find! What do the rocks on the man's desk look like?"

Herbie crept toward the desk, praying his movements wouldn't be heard by the studious man. He was wrong.

The man lifted his head and stood up, startled by the sound of something moving about in his office. He scratched his head and looked around the floor, believing his visitor to be a small critter or insect.

Herbie breathed slowly as he tried to get a glimpse of the stones. The gems were very dirty, probably recently mined and delivered to the man for cleaning. But just when Herbie was about to give up, he noticed something.

"They appear to be a deep blue and about the size of my mom's jewelry," he reported into the air.

Herbie strained his neck to look for more clues as the heat became stronger and stronger. He waited for the hat to transport him to another location, but more seconds passed as Herbie remained in the cold, dark room. Herbie clenched his fists as he felt the magic radiate against his scalp until he simply couldn't take it anymore.

Herbie stumbled to the floor of Room 13. His friends and teacher jumped at the sound of his thump. No one expected Herbie's return to be quite so soon.

Herbie frowned as he noticed his friends' confusion. What if removing the hat had ruined the king's plans for Herbie? What if he ruined this secret mission? Herbie decided the only thing to do was relay the information he gathered to his friends and teacher.

"OK, let's get right to it!" Herbie said confidently, as he ran to Ms. Pebbles' desk. "I was in a factory and then in a room with an old guy."

Marabel smiled as she read through Herbie's

observations. In the short couple of minutes that Herbie was in the factory, he managed to capture an entire system of engineering and manufacturing. Herbie's notes described large cylinder-shaped furnaces, giant cutting machines, and even diagrams of the small tools used to clean and cut gemstones in the old man's office. Everything was covered with labels and approximate measurements. It was remarkably thorough.

"So, here's the problem," Herbie began as he pointed at different pictures in his notebook. "The mineral processed in the large factory was clear, big, and rectangular in shape. But the rock that the old guy was cleaning was small and definitely blue. Is it possible we're looking for two different types of rocks?"

"Interesting, but not likely," Ms. Pebbles commented. "From everything the king and his spies know, we are only looking for one rock or mineral."

The eyes of each child widened like saucers.

Spies? What kind of mess were they in?

Marabel's eyes went back and forth from her friend's notes to the large book of minerals. She knew there was a puzzle to be solved, but she didn't know how or why. She bit her lip in annoyance.

"What's weird," Herbie's voice cut through Marabel's thoughts, "is that both Victor and I saw rocks that were blue. Gosh, this is so cool. Blue rocks!"

"That's true," Ms. Pebbles said with a spark. "Perhaps we are not just looking at minerals as they appear in nature. Perhaps we are also looking for how one could manufacture the minerals using chemistry."

"Huh?" the trio asked simultaneously.

"Let me go over this with you." Ms. Pebbles sat down at her desk, opened her large text to the first page, and began slowly reading aloud. "Rocks are made of minerals, and minerals are made of elements. Sometimes, engineers can create the minerals in labs or factories by combining different elements together. It's called a synthetic mineral. Almost like the real thing, but not quite."

A thought suddenly sprang to mind as Marabel made sense of the new information. In each of the missions, there were clues that connected together to uncover a mystery rock or mineral. But they still didn't have enough information, and the children had no idea what they were doing in the first place.

As Marabel sat and contemplated the new information, she saw it out of the corner of her eye: the blinking hardhat.

Everyone's focus shifted to the blinking hat in the corner of the room. Ms. Pebbles looked from the hat to Marabel two or three times as each of the room's occupants came to an important realization: It was Marabel's turn.

A Paxterra Pep Talk

Marabel gulped as the hat blink-blink-blinked in the corner of the room. She knew Ms. Pebbles would try to use the hat again, and she knew it would be useless. Marabel realized, perhaps before anyone else, that the hat had chosen these three children for specific reasons. Nothing is an accident, you see.

But something still bothered Marabel. She was anxious, of course, of what was to come, but she was also confused. And Marabel hated being confused.

The young girl looked at her teacher and clenched her fists. "I need to know what's going on,

Ms. Pebbles."

Ms. Pebbles shook her head. "I'm afraid there is simply not enough time, children. And, frankly, I have put you in enough danger as it is!"

Marabel sighed and reached for Ms. Pebbles' hands. "Ms. Pebbles," the girl said gently, "all three of us have already been transported in one way or another. Whether you like it or not, we are *supposed* to be involved with this."

Ms. Pebbles bit her lip. She knew Marabel was right, but she also knew that too much information could put the children at risk. She was a good teacher, after all, and her number one concern was always the well-being of her students.

Ms. Pebbles sat at her desk in deep thought. She looked up at the children crowded around her and sighed. "OK, little ones, I can tell you pieces—and only pieces—of our struggle. It is for your own good!"

Herbie, Victor, and Marabel stood like soldiers at attention as they stared at their teacher. Answers.

They were finally going to get some answers.

"A long time ago," Ms. Pebbles began, "my grandfather stumbled upon an old mine claim. After some digging within the cavern, he found an entrance into the wonderful, magical world of the Paxterras."

"The Paxterras had managed for quite some time to live unknown to the humans, and they weren't about to let a human being go out and publicize their existence! So they kidnapped my grandfather and threatened to throw him into the great fire of the mantle . . . ."

"What?" the children simultaneously shrieked.

Ms. Pebbles waved off their concern. "Oh, don't be so dramatic! It was no bother, and eventually, the great king took a liking to my grandfather. The king made the decision to learn about humans and our world, and in return, he taught my grandfather about their world as well."

"And, oh, what a world!" Ms. Pebbles exclaimed as she pointed to the posters on the wall. "Caverns, tunnels, and waterways as far as the eye can see! Oh, dear, I am getting ahead of myself. Well, as time went on, my grandfather realized what a special friendship these people had with him and that, perhaps, it should continue in our family."

Ms. Pebbles paused and looked at the ground. Her smile spread from cheek to cheek. "So the king took my father, who, as luck would have it, took me along. And a few years later, after my father died, I was given the sole responsibility of checking in on our dear Paxterra friends and making sure their needs were met."

"Their needs?" Marabel asked confused.

Ms. Pebbles nodded. "Yes, yes. From time to time, humans would bother parts of their world, perhaps mining where they should not mine, or messing with our beautiful earth in ways they should not. In any case, things were calm for the Paxterras until about two years ago when the king's trusted advisor—a stone man we know as Sulfur—turned on the king and started seeking power for himself."

The children gasped. This sounded more dangerous than they realized.

"Of course I tried to reason with Sulfur," Ms. Pebbles continued. "I have known him since I was a little girl. But he had discovered a way—through a magic far more complex and intricate than anything we have in the human world—to begin stealing minerals for himself."

"Wait, I don't understand," Marabel interrupted her teacher. "Why does it matter how many rocks or minerals they have? Isn't their entire world made of stone?"

Ms. Pebbles nodded. "Yes and no. You see, just as we use minerals in so many parts of our world, they also use minerals for their world, just in different ways. They see rocks and minerals as a great gift—the source of their very existence—so they try not to hoard too much for themselves. But that doesn't mean that, from time to time, one of the Paxterras can't be a little bit greedy, and try to take too much for themselves."

Ms. Pebbles drifted into the silence of her thoughts. She tried to look brave for her students, but she knew that the situation was dire. She took a deep breath and decided to finish the story.

"By the end of last year, Sulfur's powers grew, and he had figured out a way to steal the entire supply of a certain mineral from the Paxterra world. A rare mineral called Nicadoosite, which was unknown to humans, was lost forever. And in a matter of time, he found out how to steal from our world as well."

Ms. Pebbles sighed. "All we have are clues about which mineral he is trying to steal. We have been able to piece together the clues thus far, but it is getting trickier, and the stakes are getting higher."

Ms. Pebbles looked at her students as tears welled in her eyes. "If we only knew what mineral he is trying to steal, we could stop him. But if not, that mineral will be lost from our world and theirs forever. And I am afraid we don't have much time."

Marabel looked at Ms. Pebbles and frowned. She

hated to see grown-ups be upset, and she really wanted to help her teacher. She knew that the stakes were higher than she could have ever imagined. And yet . . . .

. . . and yet . . . fear still nearly paralyzed Marabel. She wished for Victor's daring spirit or Herbie's endless trust, but she struggled to gather up the courage to embark on other unknown adventures. And yet . . . .

. . . and yet . . . Ms. Pebbles seemed to trust Marabel and her friends more than any other grown-up ever had. The teacher really believed the magical hardhat had chosen each of the children for a special purpose. And perhaps . . . just perhaps . . . her teacher was right.

"OK," Marabel sighed as she strolled toward the hat. "I guess I'm up!"

Ms. Pebbles looked at Marabel with a surprised expression. She wiped tears from her eyes and smiled at the young girl.

Herbie rushed to his friend's side. "Wait,

Marabel! Maybe we can figure it out here in the classroom! That way you don't have to go again!"

Victor slid past Ms. Pebbles and grabbed Marabel's hand. "That's true! You're smart! And you're really good at figuring stuff out!"

Marabel shook her head. "No, guys," she said sternly. "If the hat thinks I should go, then I should go. I mean, that sounds super weird when I say it out loud, but for some reason, I think I'm supposed to go."

Ms. Pebbles smiled as she grabbed the hat from the pedestal. She stretched her arm toward Marabel and handed the young girl the blinking hardhat.

"Well done, dear girl!" Ms. Pebbles said happily. "You have discovered a great secret that took me years to uncover. When adventure calls, you must go!"

Marabel's hands shook as she stretched out her arm and clutched the hat. Victor and Herbie were both smiling with approval. Victor gave Marabel a thumbs-up gesture as Herbie mouthed "You can do it" under his breath.

Marabel pulled the hat toward her chest with a determined look, then closed her eyes. She took one more deep breath as she placed the hat on her head and, in an instant, disappeared.

The Mall and Microscopes

Swirls of sounds and people surrounded Marabel as she fell on a hard tile floor. She quickly stood and looked around. She was indoors, and there were crowds of people in every direction. She looked up and noticed dozens of stores surrounding her. She was in a shopping mall.

"Hello? And who am I speaking with?" Marabel jumped as the voice echoed through the speakers of the hardhat. She recognized the deep voice from her short time in the other world.

"Is this the king?" Marabel asked, worried she was going to knock into a busy shopper.

"Ah! You must be the adventurous spirit I met

earlier!" the king said joyfully.

"Yes, my name is Marabel." She spoke with more confidence. "Ms. Pebbles told me that I'm supposed to find a certain rock or mineral, something that links my turn with the discoveries of Victor and Herbie."

"Well, thank you for your service, Ms. Marabel! And yes, our world and yours will be in very deep trouble if we can't piece this together, and soon! Now, can you describe your location for me?"

"I'm in a shopping mall. I'm not really sure what I'm supposed to look for."

"Hmm, a shopping mall?" The king seemed confused. "I'm not familiar with this term. What do humans do in shopping malls?"

Marabel looked around at storefronts, escalators, and bright, fluorescent lights. "Well, it's where you buy stuff, like clothes and food and things like that."

"Interesting," the king spoke. "Can you see anything that would connect with what your friends saw?"

Marabel looked around. She needed to get a better look at her surroundings. She realized that because she was invisible, she could be hit at any moment by an unsuspecting shopper.

Marabel dashed toward a water fountain, hoping

to separate herself from the crowd. She stood on a ledge next to the fountain and looked around. There was a clothing store for children, a large bookstore, an ice cream parlor, and an electronics store, selling mainly cell phones.

"I'm not sure, King," Marabel spoke with frustration. "There's just so much stuff around. I don't see anything that looks like rocks or minerals!"

"No worries!" the king's voice echoed. "Remember, almost everything in your world is made from some sort of element or mineral, but it is almost always hidden to the untrained eye. Simply make notes in your mind of anything that could stand out."

Marabel started looking around at her surroundings again but quickly stopped. She grabbed at the hardhat and shrieked.

"Ouch! It feels like my head is on fire!"

"Oh, wow! Marabel, I think you know by now what this means. Get ready for another journey!"

Marabel gasped as she clutched her head and fell

toward the ground. When she opened her eyes again, she felt a soft, plush carpet beneath her feet.

"Marabel, are you there?" asked the king's voice.

The young girl slowly stood and observed her surroundings. She was in a quiet room with about a dozen other people. Pairs of men and women held hands and looked with excitement into large, glass cases.

"I'm in a jewelry store," Marabel said.

"Perfect!" the king exclaimed. "Try to get a glimpse into one of the cases!"

Marabel walked slowly behind some customers and peered over their shoulders. She looked down and saw dazzling rings, watches, and necklaces on display. Marabel bit her lip as her anxiety grew. There could be hundreds of different elements in these cases. Was she looking for gold? Silver? Copper? Diamonds? Marabel moved from case to case, each time feeling more defeated than before.

"Well, dear?" the king's voice asked with urgency.

"Nothing that stands out," Marabel responded in a low voice. Marabel paced undetected through the room as she made observations. She noticed one couple holding up what looked like an unusual engagement ring. There was a gold band, similar to her mom's ring, but instead of being clear, the stone had tints of blue. Marabel's eyes widened.

"A blue rock!" she shouted, aware for the first time that no one but the king could hear her. But before the king could answer, heat simmered through the hardhat. Marabel let out a deep, loud moan as she grasped the hat and realized she had to brace for yet another destination.

Marabel landed with a THUD on a hard tile floor. For a moment, she thought she might have returned to her classroom but was disappointed to find herself in a dark room filled with large scientific equipment. The room was occupied by only one other person: a scientist in a long, white lab coat.

"I'm in some sort of a science lab," she announced.

"Interesting," the king remarked. "Perhaps they are studying a particular rock or mineral. Try to look at what is in some of the instruments. But remember, you must not touch a thing! Any contamination could ruin their research!"

Marabel decided that with only one other person in the room it was best not to move at all. She peered as much as she could into the lens of a microscope. There was a squirming little organism floating around in a petri dish.

Shoot! Marabel became more frustrated than ever. This wasn't a rock or mineral! The scientist was studying some stupid, slimy bug!

Marabel began to panic as she looked for any clue as to what connected all three of her visited locations. It was clear that the researcher wasn't studying any element of the earth. Marabel was about to remove her helmet in defeat when she had an idea. What about the instruments?

"Oh, dear me!" the king's voice echoed through the hardhat with worry and concern.

"What's wrong, King?" Marabel asked, worried that perhaps she had been noticed at her new location.

"Marabel, I'm afraid we must cut your time short. The matter has become most urgent."

"The matter?" Marabel kept looking around for more clues. "You mean with the mystery element that needs saving?"

"Yes, precisely," the king said, annoyed. "It's time, Marabel. We must figure out the mineral *now*!"

The Corundum Conundrum

Marabel looked one more time around the room before lifting the magical hat from her head. She landed with a loud thump on the cold, hard surface of Room 13. Victor and Herbie ran to her side, lifting her to her feet and asking dozens of questions.

But Marabel wasn't listening. She kept trying to piece together her different locations. They all seemed so random. What could be the missing link?

Ms. Pebbles ushered the children toward her desk.

"OK, OK, everyone. Now, let's see. Marabel, where did the hat send you?"

Marabel described her observations to the group. She looked at Ms. Pebbles, then at her friends. She had no answers.

Ms. Pebbles smiled as she lowered her gaze and stared directly at Marabel. She placed her hands on the young girl's shoulders and gently said, "Listen, little one. You are lost in the insecurities of your mind. Do not try to overcomplicate things, Marabel. Think: Was there one thing—just one thing—that stood out to your instincts?"

As Marabel looked at Ms. Pebbles, the young girl noticed something. Around Ms. Pebbles' neck was a beautiful necklace: a thick, brown cord containing a single stone. The stone was blue and incredibly shiny, sparkling in the bright lights of the classroom.

Marabel gasped. "Ms. Pebbles! What is that stone on your necklace?"

Ms. Pebbles looked down and smiled at her gem. "Ah, this, yes. I'm surprised you just now noticed it. I suppose it is usually under my shirt so I don't lose it. This belonged to my father. Isn't it lovely? He found it while out on one of his mine claims and had the gemstone finished for me. I never take it off."

"What kind of stone is it?" Marabel asked excitedly.

"It's a sapphire. It is supposed to remind me of our Paxterra friends."

Marabel leapt in the air with excitement. She pushed past her friends and scanned the books on Ms. Pebbles' desk. After about a minute of searching, Marabel landed on a page and began skimming its contents. She grabbed the binding and started laughing.

"What? What is it?" Ms. Pebbles asked as she dashed to her pupil's side.

Marabel smiled. "It's sapphire, Ms. Pebbles! It's sapphire!"

"Sapphire?" Ms. Pebbles grabbed the book from the desk and began reading the page Marabel had marked.

"Think about it, guys. The one thing connecting

each of our visits was a blue gemstone, just like Ms. Pebbles' necklace. It has to be sapphire!"

"Well, perhaps. But there are many blue gemstones," Ms. Pebbles said.

"Yeah, but what about my trip to the mine?" Victor asked.

"Or my time in the factory?" Herbie spoke out.

Marabel smiled. "I thought about that too. But remember how Ms. Pebbles said you can

manufacture the mineral in a factory? Guys, look at the page!"

Victor, Herbie, and Ms. Pebbles looked over Marabel's shoulder and followed along as she read, "Synthetic sapphire can be produced in a factory and is incredibly durable." Marabel explained, "It says here that it's the second hardest mineral next to diamond. Herbie, didn't you say the material was hard to scratch?"

Herbie nodded. "It was also very thick!"

Marabel looked at Ms. Pebbles. "Ms. Pebbles, what elements go into making synthetic sapphire? Are there any that could be found in a tropical climate?"

Ms. Pebbles nodded. "Why, yes! There are many different elements, but . . . ."

Ms. Pebbles paused and grabbed the book from Marabel's fingers. She flipped fervently through its pages.

"Yes! There is a mineral called bauxite: a tan, soft rock found in tropical climates. But when refined, bauxite becomes a compound called alumina."

"Whoa! Wait a minute!" Victor interjected. "Alumina? Like aluminum cans?"

Ms. Pebbles nodded. "That's correct, Victor. Alumina can either be refined into sheets of pure metal aluminum, which is used to make things like aluminum cans, or it can be formed into hard crystal forms of synthetic sapphire. Oh, this is just wonderful, Marabel! Please continue!"

Marabel paced around the room as she tried keeping up with her thoughts. "Then came my mission. It's pretty obvious how it is found in a jewelry store. I even saw a couple buying a sapphire engagement ring! And in the science lab, the man wasn't studying rocks; he was studying some gross, squirmy bugs or something."

Marabel smiled at Victor, who really loved slimy bugs. "I finally figured out that it wasn't *what* he was studying, but *how* he was studying. The equipment! The equipment had to be made with something, correct? And right here in your book it says that synthetic sapphire is used in high-end optical lenses, like those found in microscopes!"

"What about the mall?" Victor asked.

Marabel nodded. "I was stumped about that, too, until I saw a place where they sell electronics. Wouldn't an electronic store be just the place to buy those new, state of the art, digital watches?"

Ms. Pebbles' eyes widened as she realized what Marabel had pieced together. "Yes! Digital watches! How brilliant you are! So very brilliant!"

"Digital watches?" Herbie and Victor asked simultaneously.

"Yeah, you guys have seen those cool commercials!" Marabel insisted. "They're watches that are also cell phones, but their screens have to be super hard so they don't crack and break. That's where sapphire comes in! Plus, I remember my cousin getting one for Christmas last year, and I think he said something about the screen being made of sapphire."

"Yes," Ms. Pebbles nodded as she pulled her cell

phone from her pocket. "Synthetic sapphire can be used in cell phone camera lenses as well as watch screens to make them more durable."

Herbie's eyes lit up. "Yeah! The small piece I knocked to the ground was about the width of a watch screen. I bet they were going to slice it into thinner pieces!"

Ms. Pebbles nearly jumped in the air at the discoveries of the children. "Brilliant! Just brilliant!"

"Wait, wait, wait!" Victor interjected into the excitement. "That still doesn't explain the blue stones we each saw on our adventures!"

Ms. Pebbles nodded and smiled. "Yes, that is a bit confusing. Minerals, you see, can now be created synthetically in a factory. But mineral compounds are also formed naturally as gemstones. Now, sapphire is the gemstone of the mineral corundum, or as my father used to say, 'What a corundum conundrum!'"

The students blink-blinked as they stared blankly at their teacher.

"Children nowadays have no sense of humor! Anyway, the sapphire gemstone can be almost every color of the rainbow. Beautiful blues, purples, and greens. Oh, but not red; red corundum is actually ruby! My favorite gemstone, in fact."

Ms. Pebbles paused and laid her hands on her desk. "Regardless, Marabel is right. Sapphire is the one great link to all of our adventures this afternoon."

The children smiled at one another then looked at their teacher. Ms. Pebbles stood still and quietly at her desk, biting her lip and moving her eyes from side to side. She was thinking of her next move.

After several seconds of silence, Marabel spoke up. "Ms. Pebbles? What now?"

Ms. Pebbles glanced at the hardhat, which had been continuously blinking since Marabel's return. The teacher clutched at the stone around her neck and sighed as she reached for the hat.

"It is my turn now." Ms. Pebbles nervously announced. "And I know what I must do."

Journeying Together

Marabel, Victor, and Herbie gasped as their teacher disappeared into the magical world. They froze in place, each wondering what they would do should their detention teacher not return.

Victor was the first to sit down at his desk. Marabel and Herbie soon followed, and each sat in silence as they anxiously awaited their teacher's return.

The tick-tick of the clock seemed to shake the walls. The tap-tap of Victor's pencil against the side of his desk echoed through the room's walls. And with each passing second, Marabel's heart thump-thumped as a lump formed in her throat.

"Maybe we should call for help?" Herbie suggested.

"And tell them what?" Victor scoffed. "That our teacher is battling a person made of stone in a magical world as she tries to save that world and ours from losing the entire supply of the crystal sapphire? Is that what you mean?"

Herbie lowered his head. "I don't know," he muttered quietly into his hands. "It was just an idea."

Marabel put her hand on Herbie's shoulder and glared at Victor. "Enough, Victor. Mocking each other isn't going to help right now."

Victor looked down and kicked at the floor below. "Sorry, Herbie."

Marabel looked at the clock and sighed. "Guys, we just have to wait."

And so they did. The clock tick-ticked and the kids tap-tapped as minute after minute passed without the return of Ms. Pebbles. Marabel felt her nerves flow into her fingertips as she tried to focus on other topics.

Marabel stood to stretch. As she reached her arms toward the ceiling, a burst of wind flowed through the classroom. The children held their breath as they grasped at the corners of their desks.

"Well, that was quite interesting!"

Herbie leapt to his feet. "Ms. Pebbles!"

Marabel and Herbie ran to their teacher. Victor stood and breathed a sigh of relief.

Marabel hugged Ms. Pebbles and pulled back. "So, did you do it? Did you stop Sulfur? Did you save the sapphire?"

As Marabel asked her questions, her gaze focused on the plain, brown cord around Ms. Pebbles' neck. The stone was gone.

"Oh, no! What happened?" Marabel asked, fearing they had been too late.

Ms. Pebbles smiled as she looked down at her empty necklace. "It is no problem. Your questions will soon be answered, but first, there is someone you must meet!"

Marabel, Victor, and Herbie looked at one another, then back at their teacher. Marabel gulped as Herbie tap-tapped the end of his pencil nervously into his journal. The children knew what they must do: It was time to meet the great king.

Ms. Pebbles waved her hands as she ushered the kids to the corner of the room. "Come now, children. We mustn't keep him waiting."

Victor skipped out of his seat and sprinted to Ms. Pebbles' side, smiling and waving his hands in the air with excitement.

Herbie, in his haste, dropped his notebook and joined Victor, sharing in his friend's excitement.

Marabel stood still at the corner of her desk. A drop of sweat moved slowly down her forehead and onto her nose. Her cheeks flushed as she bit her lip and prepared for another journey into the unknown.

Ms. Pebbles noticed Marabel's reservation. She gently placed the hardhat on the floor and moved calmly toward the young girl.

"Marabel, dear," Ms. Pebbles spoke as she

kneeled to meet the girl's gaze, "I remember a time when I was nervous to visit the world of the Paxterras. But I realized something very important in my young days: Adventure is not reserved for those who are excited about it. Adventure is for everyone! Do not let it slip by you!"

Marabel lifted her head and smiled. She looked at her teacher—frizzy gray hair, smudged lipstick, and oversized clothes. She was unique and fearless, which were the very qualities Marabel always wanted for herself. Maybe, she thought, this great experience would help her be all that she dreamed she could be.

"OK, Ms. Pebbles," Marabel said with her head high. "Let's go meet this king!"

The boys cheered in the corner as Ms. Pebbles and Marabel rushed to join them. Ms. Pebbles grabbed the hardhat from the ground and quickly brushed the dust off the light with her sleeve.

Ms. Pebbles said quickly, "OK, if we all hold hands we can be transported together. But please do not let go; the last thing I want is for you to be lost between worlds. That would put us in quite the

pickle!"

Herbie gulped as he reached for his friends'

hands: Marabel on his left, and Victor on his right. Ms. Pebbles took Marabel's left hand and looked at the children one last time, making sure no one would be forgotten.

She slowly raised the hardhat in the air as each person took a deep breath and instantly disappeared.

The Paxterra World

Ms. Pebbles, Marabel, Herbie, and Victor landed on the moist ground of the cave. Marabel's heart raced as she looked at her surroundings, remembering how scared she was on her first trip to the magical world. She took a deep breath and looked at her friends.

The four humans were in a dark corner of an underground cave, surrounded on all sides by large boulder walls. Over her right shoulder, Marabel noticed a small, lighted opening.

"Come!" Ms. Pebbles said in a hurry. "He is waiting for us, and he really hates to be kept waiting."

Ms. Pebbles hurried to the opening. There were the

sounds of whistles again, getting louder and louder with each step.

Victor covered his ears as he approached the doorway. "Geez! It is *loud*!"

Ms. Pebbles nodded as she peeked her head through the opening. "Yes, yes. They must communicate somehow with each other! You get used to the whistles after a while. Now hold on to each other! There is no 'yield to pedestrians' rule for the Paxterras."

Marabel gasped as she saw the people of stone rush past her. Their stout, gray bodies rolled over the wet soil as they zigzagged through the traffic. Victor's eyes followed one boulder after the other and smiled from ear to ear.

"OK, everyone have a hand?" Ms. Pebbles shouted over the whistles. "On the count of three, leap into the traffic! ONE . . . TWO . . . THREE!"

Ms. Pebbles quickly scurried past the rock people as the children tried their best to not stop and stare at the magical world they occupied.

Marabel's hands remained tightly entwined with her teacher's as she bit her lip with anxiety.

Ms. Pebbles seamlessly weaved through the busy traffic and whistled her own unique tune, as if she, too, were one of the Paxterra people. Finally, the group came to an abrupt stop.

Herbie, Marabel, and Victor gasped as they looked high and low. They were in the heart of the Paxterra village, and, oh, what a village it was!

Booths and shops lined the cave streets, each vendor selling rocks and minerals of all shapes, sizes, and colors. Candles floated around the high cavern ceilings without any strings or lights attached. The cavern itself was very tall, much taller than one would think for an underground chamber, and the beautiful rock formations that dangled from the ceiling provided a backdrop of crystals reflecting the candlelight.

And the people! Oh, how amazing they were! They stood as tall as Marabel, if they stood at all. Their clothing was made of small, bright rocks in almost every color, from turquoise to opal, while their skin was primarily a gray stone. Their faces were serene and gentle, and their eyes shimmered in the candlelight as if made entirely of bright, clear crystals. Small features distinguished boy and girl rocks from one another: how they wore their hair—braided diamonds for the girls; and the color of their eyes—bright red for the boys.

However, the most remarkable aspect of the Paxterras was how they traveled. At first glance, the Paxterras appeared very short, much shorter than the children. This is due to the fact that when the Paxterra people traveled long distances, they tucked their heads under their bellies and curled into balls. Their hands and feet folded underneath as their curled bodies rolled and weaved through the crowded streets of the village, gently bumping into their peers and each whistling their sweet, piercing songs.

"Come now, children!" Ms. Pebbles shouted as she adjusted the hardhat. "I know it is all rather amazing, but we just haven't the time. We mustn't keep *him* waiting!"

The threesome followed their teacher past

different shops and bakeries. Marabel smiled as she saw signs for "Rock Candy" and wondered if the Paxterras' rock candy was as delicious and sugary as the human rock candy.

The four humans made their way to the entrance of a large door covered in beautiful ruby, emerald, and sapphire stones. The door was protected by two larger Paxterras. Although they each held large spheres under their stone armor, the guards' demeanors seemed gentle, and they both smiled with excitement as they saw the new humans approaching.

Ms. Pebbles smiled as she greeted her friends. "Hello, Topaz and Quartz. So lovely to see you both."

"Hello, Ms. Pebbles," the rock sentinels spoke in unison.

"Wait, they can talk?" Herbie asked in amazement.

"Of course they can talk," Ms. Pebbles said. "In fact, they can understand just about any language in the world! It took them a while to learn to speak our language, but once they did, they caught on quite remarkably. Didn't you, gentlemen?" Ms. Pebbles turned toward the guards who smiled, showing their gleaming teeth of pearls.

"Now, may my friends and I please see the king? He is expecting us."

The two guards stepped aside and pulled a lever from the ground. A large boulder creaked and squealed as it rolled sideways, unveiling an entryway into a separate cavern. Marabel took a deep breath as she squeezed Herbie's hand. Their great adventure had landed them here: in the most important place in the entire magical world and, perhaps, the most important place for their world too.

The King's Quarters

Ms. Pebbles smiled as she hurried the children through the doorway and into a room glimmered in remarkable stonework. Like the stone people themselves, the king's quarters was meticulously put together, each stone placed in a unique spot for a special purpose.

As Marabel looked around the room, she noticed two interesting elements. First, there were picture mosaics made of small stones throughout the space, including pictures of dinosaurs and different animals. Second, no two rocks looked exactly the same.

"Are there any rocks that are duplicated, or are

they all different?" Marabel asked in a quiet voice.

Just as Ms. Pebbles started to answer, a deep voice arose from the corner of the room.

"Now that is quite the intuitive question."

The trio jumped at the sound of the familiar voice. Victor gulped and Marabel oohed and aahed as a large Paxterra, draped in a cape of bright gemstones, slowly approached the children.

Before the students could address their new friend, the king's quarters was swarmed with Paxterras of all shapes and sizes. Whistles echoed throughout the chamber as each Paxterra crowded into the room, hoping to catch a glimpse of the new visitors.

"Enough!" The king's voice commanded over the noise. The great king turned to face the children. "I apologize. My people have never seen this many humans in our world before!"

The king looked around and smiled. "And I do believe," he continued, "it may be time for them to exit. Perhaps I could have some time alone with the humans?"

The Paxterras were dismissed, and they shuffled through the space until there remained only one: the king himself.

"Hello, King Sapphire. So nice to see you again." Ms. Pebbles stepped forward and bowed as she greeted the great king. The children awkwardly followed her lead.

"Hello again, my old friend," the king said, smiling. He slowly nodded at the teacher but quickly turned his attention to the children. "And these, I am guessing, are the brave humans who helped us earlier?"

The children quickly nodded. Victor's attention changed to the marvelous mosaics throughout the king's quarters. Herbie patted his pockets, searching frantically for his notebook, only to realize he had left it in his classroom. He clenched his fists and frowned in disappointment.

But Marabel's gaze never left the king. To her, he was the most fascinating component of the new world.

The king smiled at the children and landed his gaze on Victor. "Ah, yes. Let me guess. This must be Victor, the courageous—and talkative—boy who climbed the highest mountain peaks and traversed the dangerous forests of the Caribbean!"

Victor looked at the king and smiled. "Yep! That's me!" Victor said proudly, as he pointed his thumb at his chest.

"Yes, I knew as much," the king said as he slowly moved toward Herbie. "And I am assuming this is the great observer, the boy with attention to each and every detail?"

"Y—yes, sir. I mean, Your Majesty," Herbie acknowledged nervously as he adjusted his glasses and dropped his notebook.

The king nodded and smiled as he faced Marabel. "And you, dear. You are the heroine of today's great phenomenon."

Marabel's eyes widened as she looked at the great king. "Me? A hero? No, sir. I'm afraid you must be mistaken."

The king shook his head as he approached the astonished girl. "No, no, my dear! I am a king, remember! I am hardly ever mistaken! And you, with your problem-solving skills and thoughtful observations, have saved us all from a very difficult and dangerous predicament."

The children looked at each other, then at Ms. Pebbles, who grinned and beamed at them. The king made his way toward one of the sunken walls of his quarters.

The king pointed to stones embedded in the room's walls. "You asked about these stones, Marabel. And you are correct: There is not a single stone duplicated in the entire world. Each rock and mineral has a different composition, shape, and color from the other. Isn't that remarkable? No two stones are the same, yet each seems to fit in just perfectly."

Marabel smiled, admiring the beautiful room once more. The entire land seemed like a great dream, put together by an incredibly creative and thoughtful artist. She felt lighter as her comfort increased with each simple stone.

The king slowly walked on his two small stone feet to the corner of the room. "Children, I am aware that we are short on time, but please accept my sincere gratitude for what you did today."

"I know that each of you used your unique gifts and talents to stop Sulfur from doing something very catastrophic. As a small token of my gratitude, I have something for each of you."

The king tap-tapped with his wooden cane on a bright, blue stone embedded in the wall. Topaz and Quartz rushed into the space, each carrying trays in their large, stone hands. The king smiled as his guards approached and stood on either side of him.

Victor, Marabel, and Herbie were focused on the trays and the objects which the guards carried. Victor pumped his fist and jumped in the air as each of the children realized what the great king had done for them. Hardhats! The children were each getting their own hardhat!

The Mineral Maniacs

The king let out a deep, bellowed laugh as he prepared to speak. "Well, little ones, your adventurous spirits, observant minds, and problem-solving skills have saved the beautiful sapphire for both of our worlds. As I said before, I am eternally grateful."

The children cheered and jumped with glee. Ms. Pebbles waved her hands in the air to quiet the children's excitement.

"Yes, children, you did well," Ms. Pebbles hushed the friends. "But our mission is not over yet. I am afraid, in fact, that things will only get worse."

Marabel's body froze. She stared at her teacher

then back at King Sapphire. "What do you mean?"

Ms. Pebbles sighed. "Unfortunately, Sulfur will only get stronger. He will not stop trying to steal rocks and minerals from his world . . . and yours."

The friends looked at each other and frowned. Herbie removed his glasses and scratched his head. "So there's no way to stop him?"

The king shook his head furiously. "No, no! Of course there is a way to stop him, and I have become convinced that you young maniacs may be just the people to do it."

Marabel looked up at the king in surprise. "What did you call us?"

The king smiled. "Maniacs. I hear that is a name you are often called in the human world, and perhaps not in the best way. But I believe that anything can change for the better, so why not become maniacs in pursuit of good? You can become maniacs who care for all the rocks and minerals our great earth possesses!"

Marabel smiled. The Mineral Maniacs. It was

perfect!

"The king beckoned Topaz toward the visitors. "In order for you to properly embark on the journey before you, you must be prepared. These hardhats are embedded with all the magical wonder of our world—and the safety essentials of yours. From this point onward, you must never enter our world—or any underground cavern—without a proper hardhat!"

Topaz bowed toward Victor and placed the tray in front of the young boy's face. Victor excitedly pushed his bangs out of his eyes and reached for the bright blue hat, hoisting it proudly in the air.

"I believe, Victor, that there is something written in your language on the hat for you."

Victor looked inside the hat's brim. "For your bravery and adventurous spirit." The boy looked up and smiled. "Hey, that's me!"

The king laughed. "Why, yes it is! And what about you, Herbie?"

Herbie reached for his red hat and flipped it

HERBIE   MARABEL   VICTOR

upside down. "For your gifts of observation and logic."

Marabel smiled at her friends. In the short time the children had known the king, he had managed to understand their individual gifts quite perfectly.

The young girl smiled as she reached for her hat. She ran her fingers around the smooth, purple brim as she read her inscription out loud. "For your puzzle solving and leadership."

Leadership. What began as an afternoon filled

with fear and anxiety had ended with Marabel leading her friends through the most important mission of their young lives. Her eyes welled with tears as she tucked her hat under her arm and reached for Victor's hand, then Herbie's. Together, the maniacs stood before their teacher and the great king. And yet . . . .

. . . and yet . . . questions still remained. Why does Sulfur want to steal minerals? How is he able to? And where does the magic of the hardhat come from?

The king smiled as he read the curious minds of the students. "Children, I know you have many questions. And I wish I could answer all of them now. But know this: We need your help, and we cannot defeat Sulfur without you."

Marabel, Herbie, and Victor stood tall. They smiled at the king's confidence in them, then addressed their teacher.

"OK, Ms. Pebbles," Marabel asked. "Now what?"

Ms. Pebbles brushed her hair from her forehead as she deferred to King Sapphire. The king nodded his agreement and faced the children with a serious but hopeful expression. "Maniacs, it appears we have some work to do."

THE END

SPOILER ALERT:
Do not proceed to
Digging Deeper until
after reading the
entire book.

No seriously, last chance to read the book before the mystery mineral is revealed.

Digging Deeper

Well, how did you do? Did you figure out the mystery mineral alongside the maniacs? Were you able to put together the clues to uncover which mineral can be found in watches, jewelry, and even microscopes?

If not, that's ok! This is your chance to dig deeper into all the properties and cool information about the mineral and its use in everyday life.

Thanks for journeying with us into the amazing world of rocks, minerals, and mining. And remember: YOU ROCK!

CORUNDUM

CORUNDUM WAS ORIGINALLY NAMED "CORINVINDUM" IN 1725 BY JOHN WOODWARD AFTER THE SANSKRIT HINDU WORD FOR RUBY "KURUVINDA". THE CURRENT SPELLING WAS ADOPTED IN 1794.

FORMULA: AL_2O_3

Corundum is the Crystalline form of Aluminum Oxide.

HARDNESS
1 2 3 4 5 6 7 8 **9** 10

Hardness is determined by the ability of one mineral to scratch another. At a hardness of 9 out of 10 on the Moh's Hardness Scale, corundum is the second hardest natural known mineral to only diamond.

CRYSTAL SYSTEM: HEXAGONAL

The crystal for corundum has 3 equal distant axes forming a hexagon of six equal sides plus a top and bottom.

CLASS: OXIDES

oxides, aluminum metal is bonded with oxygen

SPECIFIC GRAVITY: 3.9 - 4.1 g/cm³

Specific gravity measures the density of a material. Water has a specific gravity of 1, so Corundum is 4 times heavier than water

COMMON IMPURITIES: Cr, Fe, Ti

When in pure crystalline form, naturally transparent, corundum is clear as glass. Traces of different elements can give corundum spectacular colors. Red corundum gemstones, or rubies, occur when traces of chromium make it deep red. Other elements can give a corundum gemstone any color of the rainbow, which are all called sapphires. A blue sapphire occurs when containingtraces of iron and titaniu

LUSTER
How light interacts with the minerals surface

ADAMANTINE (BRILLIANT SHINE) ....

VITREOUS (GLASSY REFLECTION) ....

PEARLY (MULTIPLE REFLECTIVE
LAYERS CAUSES SHINE LIKE A PEARL.) ....

DIAPHANEITY (TRANSPARENCY)
How light passes through the mineral

TRANSPARENT (LIGHT FULLY PASSES THROUGH)

TRANSLUCENT (LIGHT PARTIALLY PASSES THROUGH)

OPAQUE (NO LIGHT PASSES THROUGH)

WHERE IS NATURAL CORUNDUM FOUND?
Sapphires and Rubies have been found all over the world, but primarily in the USA in Montana and North Carolina; in southeast Asian countries of Burma, Cambodia, Sri Lanka, India, Afghanistan; and several parts of Africa, including Madagascar, Kenya, Tanzania, Nigeria, and Malawi.

● RUBY
◆ SAPPHIRE

COLOR : YELLOW, RED, BLUE, VIOLET, GOLDEN-BROWN

FROM ROCKS TO PRODUCTS
How is corundum formed and used?

CORUNDUM PRESENTS A PERFECT EXAMPLE OF ALL THREE PHASES OF THE ROCK CYCLE:
1. IGNEOUS: rocks formed from cooling magma.
2. METAMORPHIC: rocks changed by intense heat & pressure
3. SEDIMENTARY: rocks formed by erosion and compaction of other rocks

SYNTHETIC SAPPHIRE AND RUBY ARE ALSO USED IN JEWELRY

GEMS CUT AND POLISHED TO BE SOLD AS JEWELRY

Gem Quality Stones are Sent to be finished

Mine for Gems

Other Stones sent for industrial uses

Stake Claim for Placer (Gravel/Creekbed) Get Permits

Stake Claim for Lode (Hard Rock) Get Permits

Rock formation discovered, believed to contain Corundum

Gems found in the gravel in creek

Gems deposited in Creek beds (Placer deposit)

Host Rock eroded over millions of years

CORUNDUM FORMED

IGNEOUS FLOWS
CRYSTALS, CALLED XENOLITHS, ARE BROUGHT TO SURFACE AND SET IN PLACE INSIDE COOLING MAGMA AND LAVA OF BASALT FLOWS

IGNEOUS INTRUSIONS
CRYSTALS, CALLED PHENOCRYST, ARE FORMED WHEN MAGMA COOLS SLOWLY INSIDE THE EARTH'S CRUST

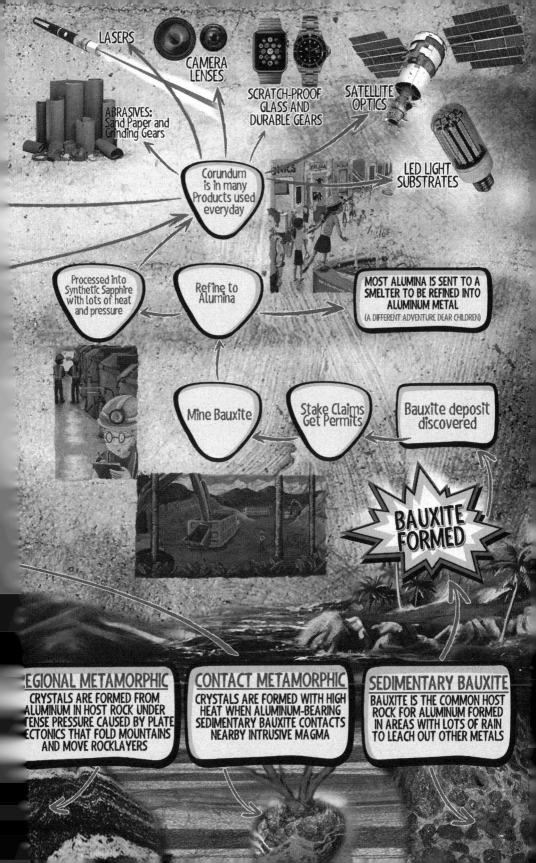

LASERS

CAMERA
LENSES

ABRASIVES:
Sand Paper and
Grinding Gears

SCRATCH-PROOF
GLASS AND
DURABLE GEARS

SATELLITE
OPTICS

LED LIGHT
SUBSTRATES

Corundum
is in many
Products used
everyday

Processed into
Synthetic Sapphire
with lots of heat
and pressure

Refine to
Alumina

MOST ALUMINA IS SENT TO A
SMELTER TO BE REFINED INTO
ALUMINUM METAL
(A DIFFERENT ADVENTURE DEAR CHILDREN)

Mine Bauxite

Stake Claims
Get Permits

Bauxite deposit
discovered

BAUXITE
FORMED

REGIONAL METAMORPHIC
CRYSTALS ARE FORMED FROM
ALUMINUM IN HOST ROCK UNDER
INTENSE PRESSURE CAUSED BY PLATE
TECTONICS THAT FOLD MOUNTAINS
AND MOVE ROCKLAYERS

CONTACT METAMORPHIC
CRYSTALS ARE FORMED WITH HIGH
HEAT WHEN ALUMINUM-BEARING
SEDIMENTARY BAUXITE CONTACTS
NEARBY INTRUSIVE MAGMA

SEDIMENTARY BAUXITE
BAUXITE IS THE COMMON HOST
ROCK FOR ALUMINUM FORMED
IN AREAS WITH LOTS OF RAIN
TO LEACH OUT OTHER METALS

WHAT IS A MINE CLAIM?

In order to explore or mine for minerals, a person must have the rights to do so. In the western United States, as established in the Mining Law of 1872, the first requirement in extracting minerals from federal lands is to "Stake a Claim".

STAKING A CLAIM:

Once discovering a site on public land that has valuable minerals, a person can mark the boundaries with stakes and then file the claim with the county clerk and the Bureau of Land Management (BLM).

TWO TYPES OF MINING CLAIMS:

Claim Stake Examples:

LODE: Minerals are in place inside the host rock

PLACER: Minerals are eroded and transported into gravel in creek

Stone Mound

3 1/2" Wood Post

2" Metal Post

WAIT A SECOND! CAN I STAKE A MINE CLAIM ON ANY LAND WITH ROCKS I WANT?

Not quite, Victor. Like most things in life, there is a process involved to claim ownership of minerals. Each place has different rules to follow so have an adult help you research the steps in your area.

Remember to obey all signs and
NEVER GO INTO AN ABANDONED MINE!

Learn more about the USA claims process at www.BLM.gov

Discover more at:

Great sources for more mineral information:

www.MineralsEducationCoalition.org

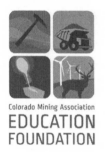

Colorado Mining Association
EDUCATION
FOUNDATION

www.AllAboutMining.org

www.usgs.gov

Acknowledgments

We first offer all glory and praise to our amazing God in Heaven, who in his faithfulness continues to light our lives and gives us the courage to follow our dreams.

This entire project began with a simple desire: to teach children about the origins of their world. There were so many people who helped us make that dream a reality. Thank you to our parents, for their endless encouragement and support. Thank you to all the friends and family who read our stories and offered technical guidance, especially Amanda Braniff, Gina Guilinger, Anita Bertisen, and many others.

Thank you to all the professionals who spent countless hours of love and care helping our story come to life. This includes first and foremost our amazing and talented illustrator, Meg Whalen, without whom we wouldn't have been able to do this! Thank you to our kind and brilliant editor, Jeaneen Dryden, who has become a dear friend and one of our biggest cheerleaders. And thank you to Gaby Seeley, our Paxterra Press Logo designer, and Jonny Black, the graphic artist for the Digging Deeper Section. You both produced beautiful work remarkably fast to help us meet our deadline. We are so grateful!

After sitting on our shelf for nearly two years, the manuscript of our story finally came to life thanks to the Move Mining competition and the hard work of its organizers. We are so grateful for your continued support. And thank you to the folks at Climax Molybdenum and The Electrum Group, who after seeing our vision come to life on the Move Mining Stage decided to step in and help make our dream a reality. We can't thank you enough for all you've done!

Thank you to my cousin, Wayne. Your life exemplified what it looks like to live fearlessly. It was your example which encouraged me to finally follow this dream without fear.

This whole project wouldn't exist without the passion, hard work, and dedication of the best man in the whole wide world. Ryan, I love you so very much, and have loved journeying with you through this crazy adventure. To many, many more!

And finally, thank you to our boys. Leo, Sam, and Max, for their patience and enthusiasm for mommy and daddy's dream. We are so proud every day to be your parents.

Author: Jules Miles

Julianne (Jules) Miles is a daughter of the Blue Ridge Mountains and was proudly raised in Charlottesville, Virginia. She has an academic background in the liberal arts, with Bachelor's and Master's degrees in Theology. She has fun in Denver, CO with her three wonderful boys and spending time with her husband, Ryan.

Illustrator: Meg Whalen

Meg studied business at Elon University in North Carolina, but soon after realized that her talents and passions were pulling her in a different direction. Meg enrolled in the children's book illustration program at Rocky Mountain College of Art & Design. While living in Denver, she also completed a master's degree in Theology at the Augustine Institute. She now lives in Florida with her husband, Danny, and their daughter, Rosie.

Made in the USA
San Bernardino, CA
16 May 2017